Arduino Byte Size
Starter Handbook and Reference

ROBERT THOMASSON

Copyright © 2013 Robert Thomasson
All rights reserved.
ISBN-10: 1492361429
ISBN-13: 978-1492361428

DEDICATION

To June and Richard.

CONTENTS

	Acknowledgments	i
	Preface	ii
1	Introduction	1
2	Light : 2.1 LED 2.2 RGB Common Anode LED	4
3	Sensors: 3.1 Tilt 3.2 Temperature 3.3 Light 3.4 Humidity	12
4	Physical Input: 4.1 Push Button Switch 4.2 4x4 Matrix Keypad	24
5	Sound: 5.1 Continuous Buzzer 5.2 Piezoelectric Speaker	29
6	Remote Control and Infra Red Detection	34
7	Introduction to Storage: SD Module	38
8	Introduction to LCD Displays: 8.1 16x2 LCD 8.2 Nokia 5110 LCD	45
9	Summary	53

ACKNOWLEDGMENTS

Thanks to the Arduino community for support and inspiration, the University of Applied Sciences Potsdam for the use of the Fritzing software tool and to my family for their endless patience.

PREFACE

There are numerous Arduino resources on the internet but sometimes it is difficult to track down specific topics. That was the exact situation that I found myself in and which ultimately drove me to write this book. I could not find a pin out diagram for a certain component so I thought that if, for the most useful and common components, there was a central reference that explained the workings of these components, showed how to connect to them to an Arduino and had a basic program to get them to work, that would be really useful. If that reference didn't get bogged down with theory, technicalities of electronics and complexities of programming, it would enable the total novice to get up and running in no time at all as well as acting as a quick reference tool when working on projects.

By example, you will learn how to get started with components as well as gaining an understanding of the basic programs to get them working. The basic understanding that you'll gain will enable you to progress to more complex projects. It's every intention not to over complicate matters; just to get to the target as quickly, and in the most uncomplicated way, as possible. Treat this book as an introduction and reference to a wide range of components and how they connect to and integrate with an Arduino. With easy instructions and break down of the code, this book provides an easy to follow handbook and handy reference.

You won't need any prior experience of electronics or programming, all you need to do is follow a few simple, logical steps. As well as the easy to follow set up instructions, most topics have a circuit diagram (all constructed using Fritzing software http://fritzing.org) and a step by step explanation of the program code.

Topics covered include light, sensors, input, sound, remote control and infra red detection and introductions to storage and displays.

The Arduino hardware and software platform is the foundation for endless developments and prototyping, the bounds of which are only limited by your imagination.

1 INTRODUCTION

Since the introduction of the Arduino open source prototyping platform, there has been a rapid increase in interest from inventors, designers, artists, engineers and above all, the hobbyist. The Arduino has opened the door to a whole new world of invention that, with just a few basic pointers, will get even a novice building their own unique prototypes.

The Arduino microcontroller boards are relatively cheap, the programming software (Integrated Development Environment; IDE) is free and can be downloaded from the Arduino website. The programming environment is written in Java and based on Processing, avr-gcc, and other open source software. The components required to build prototype systems start from just a few pence.

To follow the instructions here, most of the Arduino boards will suffice; though the book was written with the Arduino Uno Revision 3 in mind.

The first thing that you notice when you receive your new Uno is that it is so small; not as large as it appeared in the advert. Its dimensions are only 2.7 inches long and 2.1 inches wide; packed with lots of features and large enough for what we need. Normally, the Uno has a USB cable bundled with it.

The board can be powered from the DC power jack (7 - 12V), the USB connector (5V), or the VIN pin of the board (7-12V). Throughout the examples, we will only be using the USB connector which will conveniently draw the power from your computer to power the board as well as acting as an interface to upload program s and monitor output.

Next you will notice that the board has lots of pins running along the edges. All we need to know at this stage is that there are pins for power, ground, 14 digital input/output pins and 6 analog inputs. There's an ATmega328 microcontroller chip and the board has 32KB of flash memory for storing programs. That's it; all we need to know to work through all of the following examples.

As well as the USB connection, we will be using jumper wires to connect various components, usually via a breadboard to the Arduino. It is assumed that you have downloaded and installed Arduino IDE (Integrated Development Environment) from the Arduino software download page http://arduino.cc/en/main/software and have connected your Arduino to the computer using the standard USB cable (A plug to B plug).

Here is what the Arduino IDE looks like:

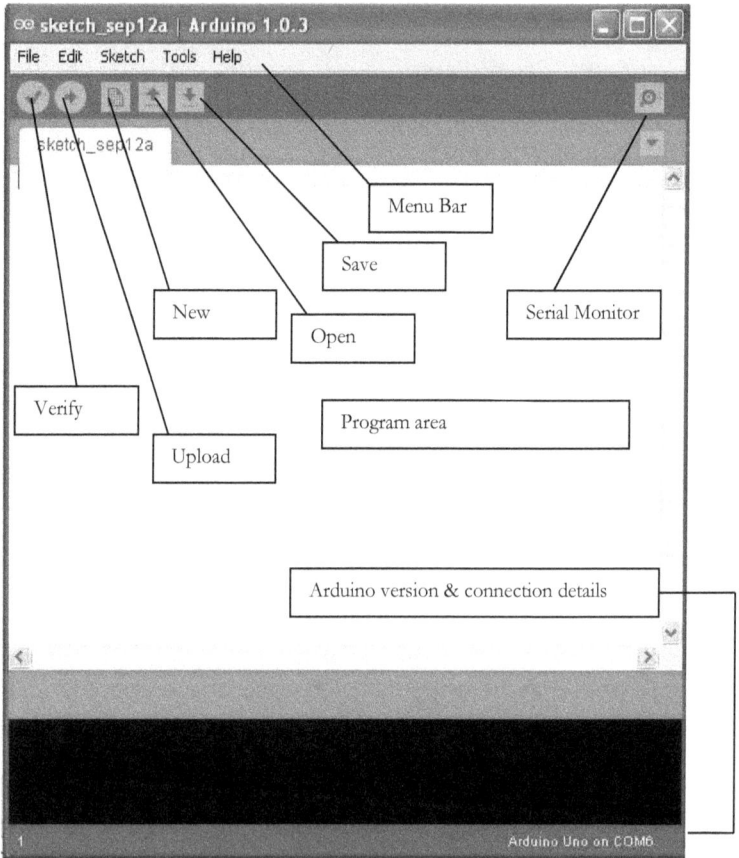

Figure 1.1 The Arduino IDE

Connect your Arduino to your computer using a USB connector and launch the Arduino IDE. Check that the correct board is selected in the IDE:

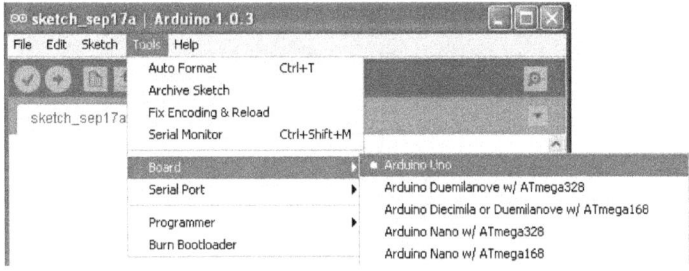

Figure 1.2 Select Correct Arduino Board

Select the correct COM port:

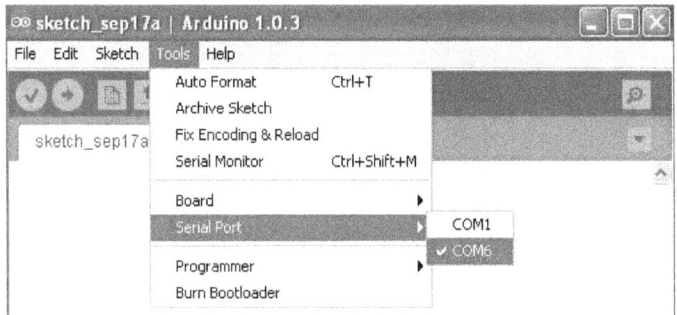

Figure 1.3 Serial Port Selection

Once you are satisfied that the settings reflect your specific Arduino board and COM port set up, you are ready to begin.

Don't get too worried about the IDE, you can keep referring back to these diagrams as and when you feel the need to. Every step of each setup is explained fully, so you shouldn't need to come back here very often.

Throughout, the example code is included in the text but it may be easier for you if you copy the code from my blog at the following address and then paste it into the Arduino IDE:

http://robertthomasson.blogspot.co.uk/2013/09/byte-size-starter.html

2 LIGHT

2.1 LED

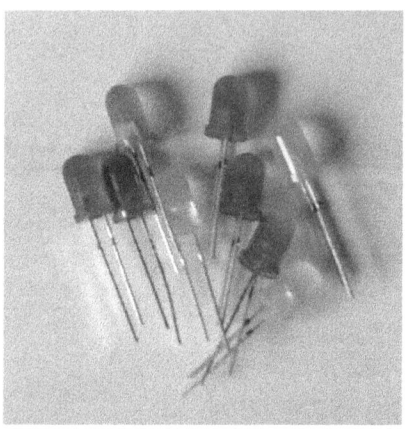

One of the most widely used electronic components is the humble light emitting diode, commonly known as LED. They are extensively used as indicator lights (for example standby lights on computers and TVs, status lights on a motorcar dashboard, advertising signs and endless other applications), are straight forward to use and are relatively cheap to buy.

Let's begin by looking at simply getting and LED to flash on and off. This example program comes with the Arduino IDE so you won't need to do any programming. Set up an LED to digital pin 13 via a 220 ohm resistor and GND on the Arduino like so:

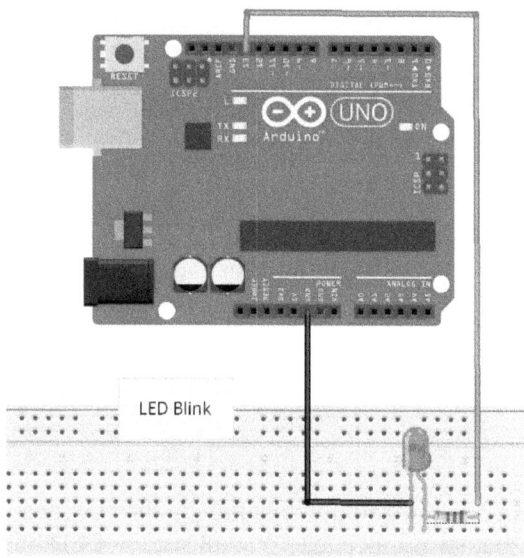

Figure 2.1 Basic Blink Circuit Diagram

Note that the longer leg of the LED (the anode) should connected to pin 13 via the 220 ohm resistor and the shorter leg (cathode) to GND. Using the drop down menus in the Arduino IDE, select: "Files", "Examples", "Basic", "Blink" to load the Blink program.

Let's take a look at the program. Firstly, you will notice lines of text beginning with "/*" and ending with "*/". These are symbols used for blocks of comments. They tell the microcontroller not to use this when executing the program. Also there is text that is preceded by two slashes "//". This is used for shorter comments; everything to the right of the two slashes on that line is a comment and will be ignored when the program is executed. These two methods of adding comments to a program are very useful and will be used in the examples that follow to explain what's going on at each step of the program.

```
/* Blink
Repeatedly turns on the LED for one second, then off for one second
*/

int led = 13;                // give the LED a name

// the setup routine runs once when you press reset:

void setup() {
```

```
pinMode (led, OUTPUT);        // initialize the digital pin as an output.

}
```

// the loop routine runs over and over again forever:

```
void loop() {
digitalWrite ( led, HIGH);    // turn the LED on (voltage level HIGH)
delay (1000);                 // wait for a second
digitalWrite (led, LOW);      // turn the LED off (voltage level LOW)
delay(1000);                  // wait for a second

}
```

Upload the program to the Arduino by clicking the "Upload" button in the IDE. Once the program is uploaded you will see the LED blinking on and off every second. The Arduino Uno board already has an LED attached to pin 13 on the board itself. If you run this example with no hardware attached, you should see that LED blink on and off.

Try altering the delay lines in the program to alter the timing of the blinking. You should note that the delay is in milliseconds so delay(1000) will give a delay of one second, delay(10000) will give a delay of ten seconds, and so on.

So far, you have connected the components to a breadboard, connected the breadboard to the Arduino and uploaded a program to the Arduino. Having successfully achieving this, the rest of this book should be relatively plain sailing.

2.2. RGB COMMON ANODE LED

The red, green, blue (RGB) common anode LED has the ability to show a multitude of colours. As the name suggests, it can be red, green or blue, or any combination of these colours. This makes it very useful if we want to show different colours by using just one LED or show an unusual colour (not red, green, or blue). This type of LED has four pins; one for each colour and one that is the common anode.

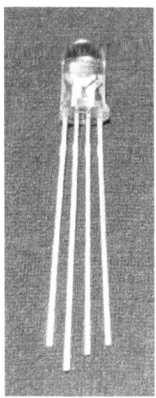

Figure 2.2 RGB Common Anode LED

Looking at the RGB LED, you will see that one of the pins is markedly longer than the other three, this is the common anode. If you hold the LED so that there is only one pin to the left of the common anode (as in the picture above), the left pin is the red cathode, the pin directly right of the common anode is the blue cathode and the pin on the extreme right is the green cathode.

Put the LED into a breadboard and wire it to the Arduino as follows:

Red pin to Arduino pin 9
Common anode pin to Arduino pin 11 via a 220 ohm resistor
Blue pin to Arduino pin 6
Green pin to Arduino pin 3

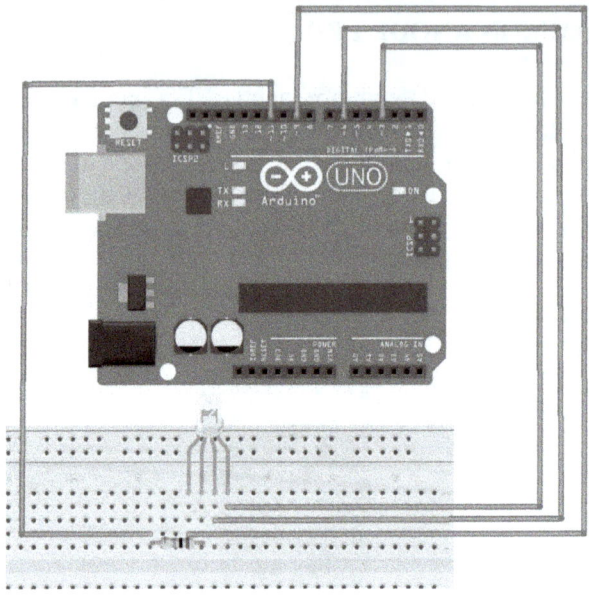

Figure 2.3 RGB Common Anode Circuit Diagram

The following program is based on code written by Dain Unicorn and EgoPlast. We don't need to go into too much detail about the code, as that would be outside of the scope of this book.

Upload the following program code (also available from http://robertthomasson.blogspot.co.uk/2013/09/byte-size-starter.html):

```
byte RedLED = 9; //red pin to Arduino pin 9
byte BluLED = 6; //blue pin to Arduino pin 6
byte GrnLED = 3; //green pin to Arduino pin 3

// define our variables...
int  RedFade;
int  BluFade;
int  GrnFade;

void setup () {
  //define the pins as output....
  pinMode (11, OUTPUT);   //common anode on Arduino pin 11
  pinMode (RedLED, OUTPUT);
  pinMode (BluLED, OUTPUT);
  pinMode (GrnLED, OUTPUT);
}
```

```
void loop () {
  digitalWrite(11,HIGH); // Enable the circuit
  analogWrite(RedLED, 255); //Turns off the RED Element
  analogWrite(GrnLED, 255); //Turns off the GREEN Element
  analogWrite(BluLED, 255); //Turns off the BLUE Element
  delay(1000);

  // Red Element fade

  for(int fadeValue = 255 ; fadeValue >= 100; fadeValue -=5) {
    analogWrite(RedLED, fadeValue);
    delay(20);
  }
  for(int fadeValue = 100 ; fadeValue <= 255; fadeValue +=5) {
    analogWrite(RedLED, fadeValue);
    delay(20);
  }

  // Green Element Fade

  for(int fadeValue = 255 ; fadeValue >= 100; fadeValue -=5) {
    analogWrite(GrnLED, fadeValue);
    delay(20);
  }
  for(int fadeValue = 100 ; fadeValue <= 255; fadeValue +=5) {
    analogWrite(GrnLED, fadeValue);
    delay(20);
  }

  // Blue Element Fade

  for(int fadeValue = 255 ; fadeValue >= 100; fadeValue -=5) {
    analogWrite(BluLED, fadeValue);
    delay(20);
  }
  for(int fadeValue = 100 ; fadeValue <= 255; fadeValue +=5) {
    analogWrite(BluLED, fadeValue);
    delay(20);
  }

  // Red+Green Elements Fade

  for(int fadeValue = 255 ; fadeValue >= 100; fadeValue -=5) {
    analogWrite(RedLED, fadeValue);
```

```
  analogWrite(GrnLED, fadeValue);
  delay(20);
}
for(int fadeValue = 100 ; fadeValue <= 255; fadeValue +=5) {
  analogWrite(RedLED, fadeValue);
  analogWrite(GrnLED, fadeValue);
  delay(20);
}
```

// Green+Blue Elements Fade

```
for(int fadeValue = 255 ; fadeValue >= 100; fadeValue -=5) {
  analogWrite(GrnLED, fadeValue);
  analogWrite(BluLED, fadeValue);
  delay(20);
}
for(int fadeValue = 100 ; fadeValue <= 255; fadeValue +=5) {
  analogWrite(GrnLED, fadeValue);
  analogWrite(BluLED, fadeValue);
  delay(20);
}
```

// Blue+Red Elements Fade

```
for(int fadeValue = 255 ; fadeValue >= 100; fadeValue -=5) {
  analogWrite(RedLED, fadeValue);
  analogWrite(BluLED, fadeValue);
  delay(20);
}
for(int fadeValue = 100 ; fadeValue <= 255; fadeValue +=5) {
  analogWrite(RedLED, fadeValue);
  analogWrite(BluLED, fadeValue);
  delay(20);
}
```

// All Elements Fade

```
for(int fadeValue = 255 ; fadeValue >= 100; fadeValue -=5) {
  analogWrite(RedLED, fadeValue);
  analogWrite(GrnLED, fadeValue);
  analogWrite(BluLED, fadeValue);
  delay(20);
}
```

```
    for(int fadeValue =100 ; fadeValue <= 255; fadeValue +=5) {
      analogWrite(RedLED, fadeValue);
      analogWrite(GrnLED, fadeValue);
      analogWrite(BluLED, fadeValue);
      delay(20);
    }

    analogWrite(RedLED,100);
    analogWrite(GrnLED,50);
    analogWrite(BluLED,0);
    delay(1000);
}
```

See how the LED fades between colours; pretty cool, eh? This program, though relatively long, is organised into distinct sections to show how to effects can be achieved. At this stage, you don't need to understand all what's going on in the program, once you have more experience, that's when you'll probably want to learn more.

3 SENSORS

A sensor, or detector, is a device that measures a physical quantity and coverts it into a signal. This signal can then be read. Sensors are very common and are all around us in everyday situations. They include devices for measuring temperature, light, motion, humidity, moisture, vibration, electrical field, magnetic field and sound.

In this section, we will investigate several of these sensors and I will illustrate the ease of their connection to an Arduino as well as using and/or manipulating their signal.

3.1 TILT

A tilt sensor can be a useful component in numerous projects. Fixed to objects, it switches between two states if moved or tilted. The SW-520D is a simple ball tilt switch which has two pins. With the two pins level, a 10° tilt will close the contacts, thus altering the "state".

As well as the sensor, we will also attach an LED so that we will have a visual indication of the sensors state. Firstly connect an LED to pin 13 and GND directly on the Arduino. The rest of the wiring is straight forward I.e. connect one of the pins of the sensor to ground and the other pin to a 10k ohm pull up resistor and to Arduino pin 6.

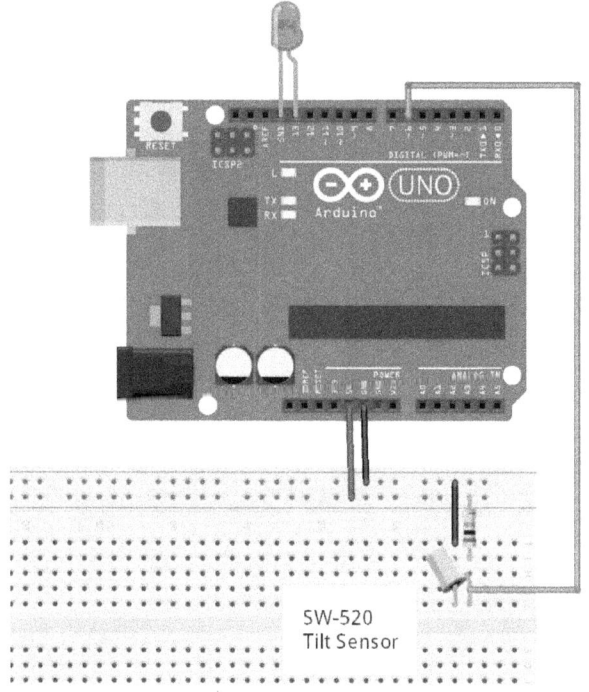

Figure 3.1 Tilt Sensor Circuit Diagram

Upload the following program (also available from http://robertthomasson.blogspot.co.uk/2013/09/byte-size-starter.html):

```
const int sensorPin = 6;        // tilt sensor pin
const int ledPin = 13;          // LED pin
int sensorState = 0;            // define our variable

  void setup() {
```

```
  Serial.begin(9600);            // enable serial output and baud rate
  pinMode(ledPin, OUTPUT);       // define LED pin as output
  pinMode(sensorPin, INPUT);     // define sensor pin as output

}

void loop(){

String s;                        // create a string variable called s
sensorState = digitalRead(sensorPin);  // state = value of the pin

if (sensorState == HIGH) {       // if it's high…..
  digitalWrite(ledPin, HIGH);    // turn on the LED
  s = "TILT";                    // value of our variable
}
else {                           //otherwise….
  digitalWrite(ledPin, LOW);     // turn off the LED
  s = "FLAT";                    // value of our variable
}

Serial.println (s);              // print the value in the Serial Monitor
delay(1000);                     // wait for 1 second

}
```

When the sensor is moved, its state (either FLAT or TILT) will be displayed in the Serial Monitor of the IDE and the light will flash correspondingly. Move the breadboard that is housing the tilt sensor to see this happening.

3.2 TEMPERATURE

Here, we are going to use an LM35 precision analog temperature sensor. It has 3-pins; 5V, GND and Temperature.

As you look at the flat face of the sensor with the pins hanging downwards, the left pin is the 5V pin and should be attached to the 5V pin of the Arduino. The right pin of the sensor is ground and should be connected to the Arduino GND pin. Finally, the central pin should be connected directly to Analog input pin 0 of the Arduino. Notice that we are not using a breadboard here. This is because if we were to use a breadboard, the temperature wouldn't be very stable. If you must use a breadboard, you would be well advised to add some sort of filter.

Figure 3.2 Temperature Sensor Circuit Diagram

Upload the following program (also available from http://robertthomasson.blogspot.co.uk/2013/09/byte-size-starter.html): and open the Serial Monitor. You should see the temperature reading value being updated every second. If you grab hold of the sensor or breathe on it, you should see the temperature rise. Try altering the delay so that you see the temperature being updated at different intervals (remember that this is milliseconds so delay(1000); pauses for one second).

```
#define tempPin 0            // sensor attached to analog pin 0

int Temp = 0;                // set up a variable

void setup() {
Serial.begin(9600);          // enable serial output and baud rate
```

```
void loop(){

analogRead(tempPin);                              // read the pin value
Temp = (5.0 * analogRead(tempPin) * 100.0) / 1024;   //see below
Serial.print("Temp ");                            // print "Temp"
Serial.print(Temp);                               // print the value
Serial.println("");                               // print a blank line
delay(1000);                                      // wait 1 second
}
```

When the Arduino is using the 5V reference with 10bit analog to digital conversion (ADC) it has a 1024 count. Therefore, we have to perform the following calculation on the sensor measurement to give true temperature in Centigrade.

Temp = (5.0 * analogRead(tempPin) * 100.0) / 1024

Look at the following lines of code:

Serial.print("Temp ");   // print the word "Temp" in the Serial Monitor
Serial.print(Temp);   // print the value in the Serial Monitor
Serial.println("");   // print a blank line in the Serial Monitor

Rather than just have a number being displayed, I have added a couple of lines so the word "Temp" is displayed before the value and " degrees centigrade" after the value. I have also included a line to print a blank line.

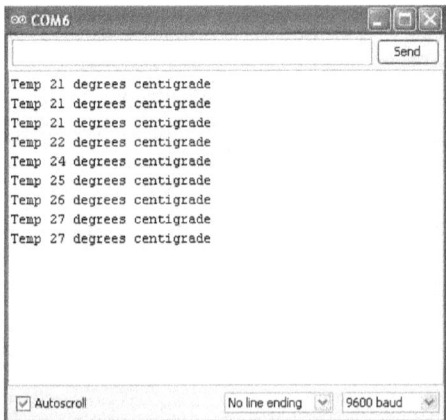

Figure 3.3 Temperature Serial Monitor Display

This makes the display look neater and easier to read. Try adding some different text or more spaces. You could also try altering the time between readings.

3.3 LIGHT

Light sensors, also known as photo resistors or light dependant resistors (LDR) are generally quite basic and uncalibrated. In this example, I have used a GL5537 LDR but there are plenty of others that perform equally as well. The LDR changes its resistance depending on the amount of light and we can measure this change using one of the Arduino analog pins.

The LDR has only two pins so connect one of these pins to GND (ground) and the other pin to a 10k ohm pull up resistor and to Arduino analog pin 0 as shown below.

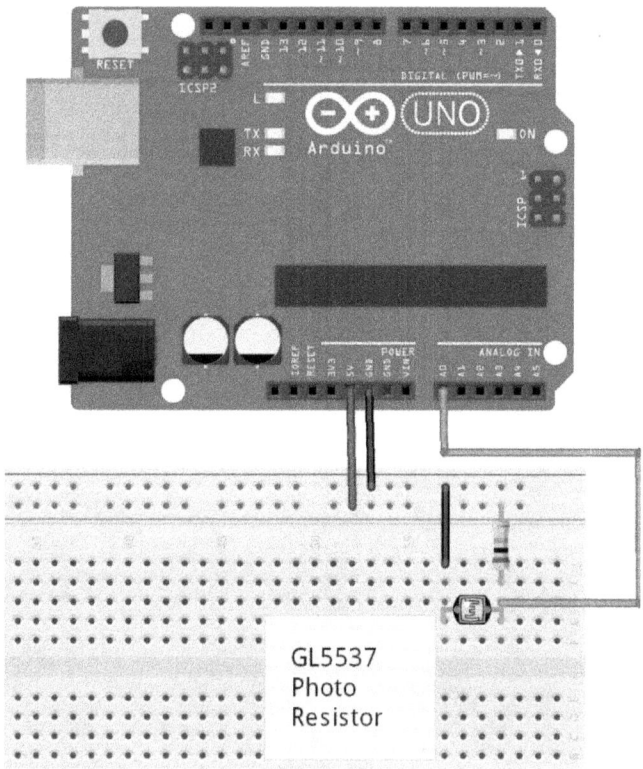

Figure 3.4 Photo Resistor Circuit Diagram

Upload the following program (also available from http://robertthomasson.blogspot.co.uk/2013/09/byte-size-starter.html) and open the Serial Monitor.

```
#define lightPin 0            // data pin
int Light = 0;                // our variable
```

```
void setup() {
  Serial.begin(9600);           // enable serial output and baud rate
}

void loop(){
  analogRead(lightPin);         // read the pin value
  Light = analogRead(lightPin); // Light variable = value that was read
  Serial.println (Light);       // Print the value in the Serial Monitor
  delay(1000);                  // Wait for 1 second
}
```

Placing your hand (or piece of paper) over the sensor results in a change of values displayed. The more light, the lower the reading.

As the measured resistance varies greatly from bright light to darkness, we need to either use a lookup table or perform calculations on the readings to convert to lux values.

The following program converts the measurements to lux (also available from http://robertthomasson.blogspot.co.uk/2013/09/byte-size-starter.html):

```
#define lightPin 0
int Light = 0;

void setup() {
  Serial.begin(9600);
}

void loop()
{
  Light = toLux(analogRead(lightPin));       //perform conversion
  Serial.println (Light);

  delay(1000);
}
// Change the ADC reading to lux.
/*         Vcc     5
           Pullup  10000
  lux      ohms    Voltage  ADC     Scaling
  0                         1023
  1        200000  4.76     975     -0.020937188
  10       30000   3.75     768     -0.043428309
  80       5000    1.67     341     -0.1640625
  120      4000    1.43     293     -0.8203125
  150      3000    1.15     236     -0.533203125
```

```
250    2000   0.83   171   -1.5234375
300    1800   0.76   156   -3.45703125
800    700    0.33   67    -5.604580966
*/

int toLux(int adc)
{
  // return (map(adc, 0, 1023, 900, 0)); simple linear model
  if (adc > 868)
    return 300 + 5.6 * (adc - 868);
  else if (adc > 853)
    return 250 + 3.46 * (adc - 853);
  else if (adc > 788)
    return 150 + 1.52 * (adc - 788);
  else if (adc > 731)
    return 120 + 0.53 * (adc - 731);
  else if (adc > 683)
    return 80 + 0.82 * (adc - 683);
  else if (adc > 256)
    return 10 + 0.16 * (adc - 256);
  else
    return 0.04 * adc;
}
```

The first program (without the conversion to lux) would be sufficient if you just want to detect the difference between light and shadow, if a light is on or off or there is just a change in brightness.

3.4 HUMIDITY

Another sensor that you may wish to use is the humidity sensor. In this example I have used the DHT-11. This is a digital humidity and temperature sensor and has four pins. The measurement resolution is 1°C or 1% relative humidity we will just use it for measuring relative humidity and measure temperature with the more accurate and sensitive LM35 (see Section b above).

Simply connect Data pin to Arduino digital input pin 7 and the indicated pins to +5V and GND as illustrated.

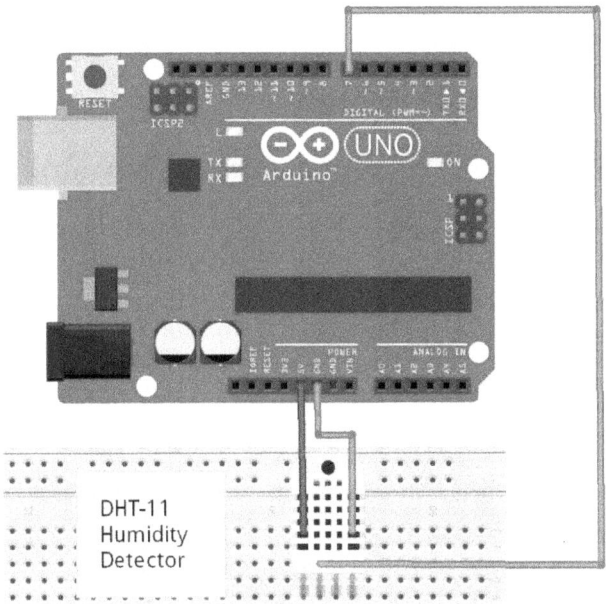

Figure 3.5 Humidity Detector Circuit Diagram

In this example we will be using a library as part of the code. Libraries provide extra functionality for use in programs and are referenced from within the program. You will need to download the DHT 11 library and install it. https://github.com/ghadjikyriacou/DHT11lib/blob/master/dht11.h

Instructions for installing Arduino libraries can be on the Arduino website found at http://arduino.cc/en/Guide/Libraries . Once the library has been installed, upload the following code (also available from http://robertthomasson.blogspot.co.uk/2013/09/byte-size-starter.html) and open the Serial Monitor:

```
#include <dht11.h> // use the dht11 library
int Humidity, hCheck;

#define humidityPin 7            // define humidity pin as pin 7
dht11 DHT11;

void setup()
{
pinMode(10, OUTPUT);
Serial.begin(9600);
}

void loop()
{
 hCheck = DHT11.read(humidityPin);
 if(hCheck != 0)
 Humidity = 255;                 //this must be an error
 else
 Humidity = DHT11.humidity;

Serial.print (Humidity);                    //print the measurement
Serial.print (" % Relative Humidity");      //print some text
Serial.println ("");                        //print a blank line
delay (1000);                               //wait for 1 second
}
```

You will see that the humidity measurement is updated in the Serial Monitor once every second.

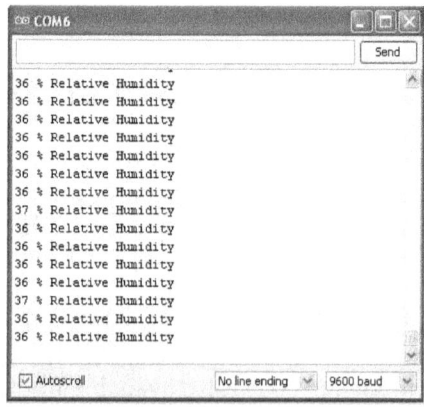

Figure 3.6 Serial Monitor Display of Humidity

If you breathe on the detector, you should see the humidity measurement change. Try altering the delay interval or the way that the data is displayed in the Serial Monitor.

This chapter has introduced some very useful sensors and hopefully inspire you to experiment with them. Try attaching more than one to the Arduino, not forgetting that you will have to amalgamate the code for each individual sensor.

4 PHYSICAL INPUTS

We are all familiar with typing on a keyboard or clicking a mouse to relay information to some sort of processor. Here we will look how we can connect a component to the Arduino, press something and then have the Arduino act accordingly. Firstly we will use the most simple of these components, the momentary tactile push button switch. Secondly we will set up a 4x4 matrix keypad.

4.1 PUSH BUTTON SWITCH

Push button switches are easy to set up and can be used in innumerable situations. When pressed, they connect two points in a circuit. This basic example turns on an LED when the button is pressed. As well as breadboard, wires and the button itself, you will need a 10k ohm resistor. Here is the circuit diagram:

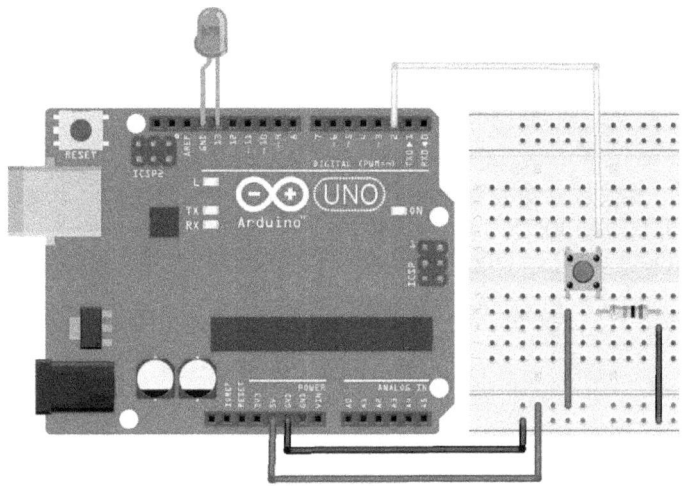

Figure 4.1 Push Button Switch and LED Circuit Diagram

Upload the following program that is based on the button example on the Arduino website at http://arduino.cc/en/tutorial/button (also available from http://robertthomasson.blogspot.co.uk/2013/09/byte-size-starter.html):

```
const int buttonPin = 2;        // the pushbutton pin
const int ledPin =  13;         // the LED pin

// variables will change:
int buttonState = 0;            // variable for reading the pushbutton status

void setup() {

  pinMode(ledPin, OUTPUT);    // initialize the LED pin as an output
  pinMode(buttonPin, INPUT);  // initialize the pushbutton pin as an input

}

void loop(){

  buttonState = digitalRead(buttonPin); // read the state of the pushbutton
```

```
 // if pushbutton, the buttonState is HIGH:
 if (buttonState == HIGH) {
 digitalWrite(ledPin, HIGH);    // turn LED on
 }

else
 {   digitalWrite(ledPin, LOW); // turn LED off
  }

}
```

This basic circuit and program can serve as part of much more complex projects. Try building a circuit that incorporates a button or more than one button with an LCD or any of the other component(s) that are covered in this book.

4.2 4x4 MATRIX KEYPAD

The 4x4 matrix keypad are very accessible and you can pick them up quite cheaply on the internet.

The arrangement of the keys is

1 2 3 A
4 5 6 B
7 8 9 C
* 0 # D

There is a ribbon with 8 wires running from the bottom of the keypad. With the keypad face up, the wires connect in sequence from left to right to Arduino digital pins 2 - 9. Don't use digital pins 0 and 1 on the Arduino Uno, since they are used for serial communication.

The Arduino Keypad library is available from the Arduino Playground.

Note: I also have an I2C version of this example (01/24/13).

The following code is based on that by http://bradsduino.blogspot.co.uk and illustrates the basic functionality. Upload the program (also available from http://robertthomasson.blogspot.co.uk/2013/09/byte-size-starter.html) and open the Serial Monitor.

```
#include <Keypad.h>

const byte ROWS = 4;
const byte COLS = 4;
```

```
char keys[ROWS][COLS] = {
 {'1','2','3','A'},
 {'4','5','6','B'},
 {'7','8','9','C'},
 {'*','0','#','D'}
};
byte rowPins[ROWS] = {2,3,4,5}; //connect to row pinouts
byte colPins[COLS] = {6,7,8,9}; //connect to column pinouts

Keypad keypad = Keypad( makeKeymap(keys), rowPins, colPins, ROWS, COLS );

void setup(){
  Serial.begin(9600);
}

void loop(){
  char key = keypad.getKey();

  if (key != NO_KEY){
    Serial.println(key);
  }
}
```

As each key is pressed, the key's corresponding character appears on a separate line in the Serial Monitor. This is very useful and can be incorporated into numerous projects. For example, press a particular key, do something; press another key, do something else. Try experimenting by incorporating different components elsewhere in this book and their programs to develop your own unique projects.

5 SOUND

Sound is all around us and it is very useful as an indicator or as a warning. Think of the beeping at a zebra crossing, a fire alarm, a police siren; all distinguishable and all memorable. We may want to use sound in our future projects so this chapter will introduce you to two very basic, but highly versatile components.

5.1 CONTINUOUS BUZZER

Figure 5.1 5V Continuous Buzzer

The continuous buzzer has only two pins so we need to connect one to GND and the other to a digital pin on the Arduino; in this example it is connected to digital pin 13.

Once you have connected the buzzer, load the following simple program (also available from http://robertthomasson.blogspot.co.uk/2013/09/byte-size-starter.html):

```
int buzzer = 13;           //define buzzer pin

void setup()
{
  pinMode(buzzer, OUTPUT);   //buzzer pin as OUTPUT
}

void loop() {

digitalWrite(buzzer, HIGH);   // make voltage HIGH, turn the LED on
digitalWrite(buzzer, LOW);    // make voltage LOW, turn the LED off
delay(1000);                  // wait for a second

}
```

This very simple program turns the buzzer on, then off, then waits for a second before repeating. Experiment by altering the delay time and putting extra delay commands into the program. For instance edit the program so that it looks like the following and see the difference of adding just one extra delay command:

```
int buzzer = 13;          //define buzzer pin

void setup()
{
 pinMode(buzzer, OUTPUT);   //buzzer pin as OUTPUT

}

void loop() {

digitalWrite(buzzer, HIGH); // turn the LED on by making the voltage HIGH
 delay(1000);            // wait for a second
digitalWrite(buzzer, LOW);  // turn the LED off by making the voltage LOW
 delay(1000);            // wait for a second

}
```

With some imagination and dedication, I'm sure that you can compose something memorable.

5.2 PIEZOELECTRIC SPEAKER

A piezoelectric speaker can play, as well as detect tones and can, with a little help, enable us to play actually melodies. By using the Arduino capability of producing pulse width modulation (PWM) signals, we are able to alter the pitch of the sound. If you need to know more about PWM, details can be found here: http://webzone.k3.mah.se/k3dacu/projects/ivrea/motor/pwm.html

Usually, the two wires of the piezoelectric speaker are coloured. The black wire should be connected to GND and the red wire connected to the output (in this case, digital pin 10 of the Arduino).

Here is a very basic program that will play a single note (also available from http://robertthomasson.blogspot.co.uk/2013/09/byte-size-starter.html):

```
const int speakerPin =10;            // speaker is on digital pin 10

void setup() {

pinMode(speakerPin, OUTPUT);

}

void loop() {

int speakerVal = 349;                //input speakerVal to alter the note
//the value of 349 corresponds to the note F (see bleow)
tone(speakerPin, speakerVal);        // calculate and play the tone

}
```

The required note frequencies to compose melodies using the set up can be found at http://www.phy.mtu.edu/~suits/notefreqs.html , a summary of which is:

C 262 Hz
C# 277 Hz //middle C
D 294 Hz
D# 311 Hz
E 330 Hz
F 349 Hz
F# 370 Hz
G 392 Hz
G# 415 Hz
A 440 Hz

A# 466 Hz
B 493 Hz
C 523 Hz
C# 554 Hz
D 587 Hz
D# 622 Hz

Try changing the note that is played and try adding extra lines of code to add more notes. By adding various delay commands and different notes you should be able to compose a simple melody. I'm sure that you will have hours of fun with this, whilst also gaining valuable experience by working with the program.

6 REMOTE CONTROL AND INFRA RED DETECTION

There a several infrared sensors on the market but I have found that the HX1836 is a versatile, reliable and inexpensive universal detector. They are easy to set up and before long you'll have everything working at the push of a button. Lately, I've see more and more "infra red modules" appearing on the market. These consist of an HX1836 detector mounted on a board and a small NEC remote control handset. If you only have the detector, don't worry as most remote control handsets that you might find lying around should be fine for this. In the past, I have used an old satellite TV handset that was about to be put in the bin.

Firstly, we'll wire up the infrared receiver then get the codes from an infrared remote control handset.

Fig 6.1. HX1836 Universal Detector

As you look at the X on the face of the detector, the pins are –

Right   VCC (5V)
Middle   GND
Left   IN   Attach this to digital pin 9 on the Arduino

Firstly, carefully put the pins of the detector into a breadboard and run jumper wires from these to the relevant Arduino pins. If your detector is on a board, you can wire directly from the board pins to the Arduino.
Once wired up, we need to see if it's working properly. So upload the following program to the Arduino (also available from http://robertthomasson.blogspot.co.uk/2013/09/byte-size-starter.html):

```
#include <IRremote.h>                        // use the IRremote library

int RECV_PIN = 9;                            // IR receive pin is digital pin 9
IRrecv irrecv(RECV_PIN);
decode_results results;

void setup()
{
  Serial.begin(9600);                        // enable serial output and baud rate
  irrecv.enableIRIn();                       // Start the receiver
}

void loop() {
  if (irrecv.decode(&results)) {
    //irrecv.resume();
    Serial.println(results.value, DEC);      // display value in Serial Monitor
    irrecv.resume();                         // Receive the next value
  }
}
```

If we now open the Serial Monitor and start pressing the buttons on the handset, you will see the corresponding button codes displayed:

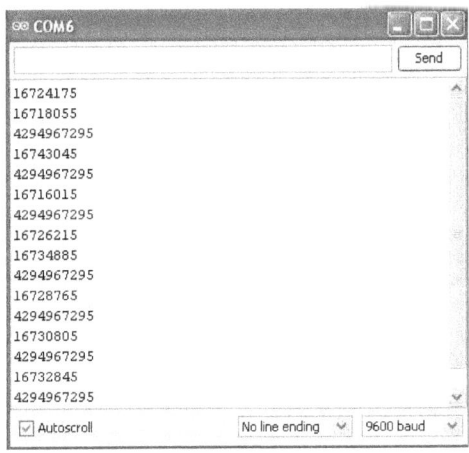

Fig 6.2 Serial Monitor Display: Infrared Remote Control Handset Codes

If you wish to use the handset in a project, you will need to make a note of the codes that correspond to the keys that you wish to use. For example, if you want to turn on an LED by pressing "1" on the keypad, you will have to use the value 16724175 in your program.

Here are the key codes that I got from the NEC remote control handset.

CH-	CH	CH+
16753245	16736925	16769565
<<	>>	>
16720605	16712445	16761405
-	+	EQ
16769055	16754775	16748655
0	100+	200+
16738455	16750695	16756815
1	2	3
16724175	16718055	16743045
4	5	6
16716015	16726215	16734885
7	8	9
16728765	16730805	16732845

If you add an LED to digital pin 13 as in the earlier chapter on LEDs, you can turn it on and off by pressing the "1" key using the program below (also available from http://robertthomasson.blogspot.co.uk/2013/09/byte-size-starter.html):

```
#include <IRremote.h>          //use the IRremote library
int led = 13;                  //LED on pin 13
int RECV_PIN = 9;              //detector on pin 9
IRrecv irrecv(RECV_PIN);
decode_results results;

void setup()
{pinMode(led, OUTPUT);         //LED pin as output
  Serial.begin(9600);
  irrecv.enableIRIn();         // Start the receiver
}

void loop() {
```

```
  digitalWrite(led, LOW);           //LED off
  if (irrecv.decode(&results)) {
//irrecv.resume();
    Serial.println(results.value);
  irrecv.resume(); // Receive the next value
  }

  if (results.value == 16724175)    //if button 1 pressed...
  {
  digitalWrite(led, HIGH);          //turn on LED
  delay (2000);                     //keep LED on for 2 seconds
  digitalWrite(led, LOW);           //turn off LED
  }

  else digitalWrite(led, LOW);

}
```

Try experimenting with more than one LED or, better still, change the colour of an RGB common anode LED by pressing different buttons on the handset. These and much, much more will be covered in depth in future publications.

7 INTRODUCTION TO STORAGE: SD MODULE

Earlier whilst we were looking at sensors, we were generating a lot of data. This data, though displayed in the Serial Monitor, has been lost forever as we had no means of saving it. If you are logging data, the chances are that you will want to save it for later analysis. One of the easiest ways to store data for the sensors is the familiar SD storage card. SD Card Modules can be easily found and are quite reasonably priced.

Figure 7.1 SD Card Module

The SD Card Module has eight pins, which are all labelled. The pin out scheme that I use is as follows:

SD Card	Arduino
GND	GND
MISO	12
SCK	13
MOSI	11
CS	10
5V	
3V	3V
IRQ	

Note that the SD Card Module can be used with either 5V or 3V connected to the Arduino. The following programs have been adapted from the examples that are freely available in the Examples Section of the Arduino IDE. You will need to ensure that you have the SD Card library ("SD.h") installed.

7.1 List The Contents Of An SD Card

Firstly, save some files to the SD card using your computer our laptop. We will now upload the following program (based on the "CardInfo" file from the Arduino IDE(also available from http://robertthomasson.blogspot.co.uk/2013/09/byte-size-starter.html):

```
#include <SD.h> // include the SD library:

// set up variables using the SD utility library functions:
Sd2Card card;
SdVolume volume;
SdFile root;

// change this to match your SD shield or module
// most common SD shields and modules: pin 10

const int chipSelect = 10;

void setup()
{
  Serial.begin(9600); // Open serial communications and wait for port to open:
  while (!Serial) {
    ; // wait for serial port to connect.
  }
  Serial.print("\nInitializing SD card...");
  pinMode(10, OUTPUT);    // change this to 53 on a mega

  // we'll use the initialization code from the utility libraries
  if (!card.init(SPI_HALF_SPEED, chipSelect)) {
    Serial.println("initialization failed. Things to check:");
    Serial.println("* is a card is inserted?");
    Serial.println("* Is your wiring correct?");
    Serial.println("* did you change the chipSelect pin to match your shield or module?");
    return;
  }
  else {
    Serial.println("Wiring is correct and a card is present.");
  }

  // print the type of card
  Serial.print("\nCard type: ");
  switch(card.type()) {
```

```
    case SD_CARD_TYPE_SD1:
      Serial.println("SD1");
      break;
    case SD_CARD_TYPE_SD2:
      Serial.println("SD2");
      break;
    case SD_CARD_TYPE_SDHC:
      Serial.println("SDHC");
      break;
    default:
      Serial.println("Unknown");
  }

  // Now we will try to open the 'volume'/'partition' - it should be FAT16 or FAT32
  if (!volume.init(card)) {
    Serial.println("Could not find FAT16/FAT32 partition.\nMake sure you've formatted the card");
    return;
  }

  // print the type and size of the first FAT-type volume
  uint32_t volumesize;
  Serial.print("\nVolume type is FAT");
  Serial.println(volume.fatType(), DEC);
  Serial.println();

  volumesize = volume.blocksPerCluster();    // clusters are collections of blocks
  volumesize *= volume.clusterCount();       // we'll have a lot of clusters
  volumesize *= 512;                         // SD card blocks are always 512 bytes
  Serial.print("Volume size (bytes): ");
  Serial.println(volumesize);
  Serial.print("Volume size (Kbytes): ");
  volumesize /= 1024;
  Serial.println(volumesize);
  Serial.print("Volume size (Mbytes): ");
  volumesize /= 1024;
  Serial.println(volumesize);

  Serial.println("\nFiles found on the card (name, date and size in bytes): ");
  root.openRoot(volume);

  // list all files in the card with date and size
```

```
    root.ls(LS_R | LS_DATE | LS_SIZE);
}
```

```
void loop(void) {

}
```

If you now open the Serial Monitor, you will see a list of all files in the card with their date and size.

7.2 Write To An SD Card

To write to a file, we firstly need to create that file on the SD card. Open up a new file using Microsoft Notepad or similar, then save it as "test.txt" to the SD card. Upload the following program (also available from http://robertthomasson.blogspot.co.uk/2013/09/byte-size-starter.html):

```
#include <SD.h>   // use the SD library

File myFile;         // define the file

void setup()
{
Serial.begin(9600);
Serial.print("Initializing SD card...");
pinMode(10, OUTPUT);

if (!SD.begin(10)) {
Serial.println("initialization failed!");
return;
}
Serial.println("initialization done.");

// open the file. Note that only one file can be open at a time,
// so you have to close this one before opening another.

myFile = SD.open("test.txt", FILE_WRITE);

if (myFile) {
// if the file opened, write to it:
Serial.print("Writing to test.txt...");
myFile.println("testing 1, 2, 3.");
myFile.close();              // close the file:

Serial.println("done.");

}

else {
// if the file didn't open, print an error:
Serial.println("error opening test.txt");
}

}
void loop()
```

```
{
// nothing happens after setup
}
```

If you put your SD card in the SD card slot of your computer and open the file "test.txt", you will see that the Arduino has written the text "testing 1, 2, 3." to the file. This is very useful as this program can be used for the basis of a project. For example, you could set up a temperature sensor and log the data to an SD card for future manipulation and analysis.

8 INTRODUCTION TO LCD DISPLAYS

In earlier chapters we have looked at data displayed in the Serial Monitor and we have also managed to store data on an SD card. What if we want to display data that is generated on a stand-alone Arduino that isn't connected to a computer? The answer is that we need some sort of display. There are numerous sizes and types available but we will be concentrating on two of the most common of the liquid crystal display (LCDs).

8.1 16x2 LCD Display

The 16x2 LCD display is very common and can be used in numerous situations and projects. They come with a variety of text and background colours, so pick the one that best suits the situation or your taste. These LCDs either come with header pins already attached, or with header pins that you have to solder on yourself. There are also an increasing number of outlets selling these LCDs as a shield with the LCD connected to a board which has buttons connected; the whole thing having the ability to just slot onto the Arduino. Here, we will assume that your LCD has the pins already attached.

Firstly, place the LCD onto a breadboard, leaving enough room for a 10k ohm potentiometer (variable resistor). This is used to brighten or dim the display on the LCD. You will also need a 220 ohm resistor. Now, wire up the pins of the LCD to the Arduino as in the diagram below:

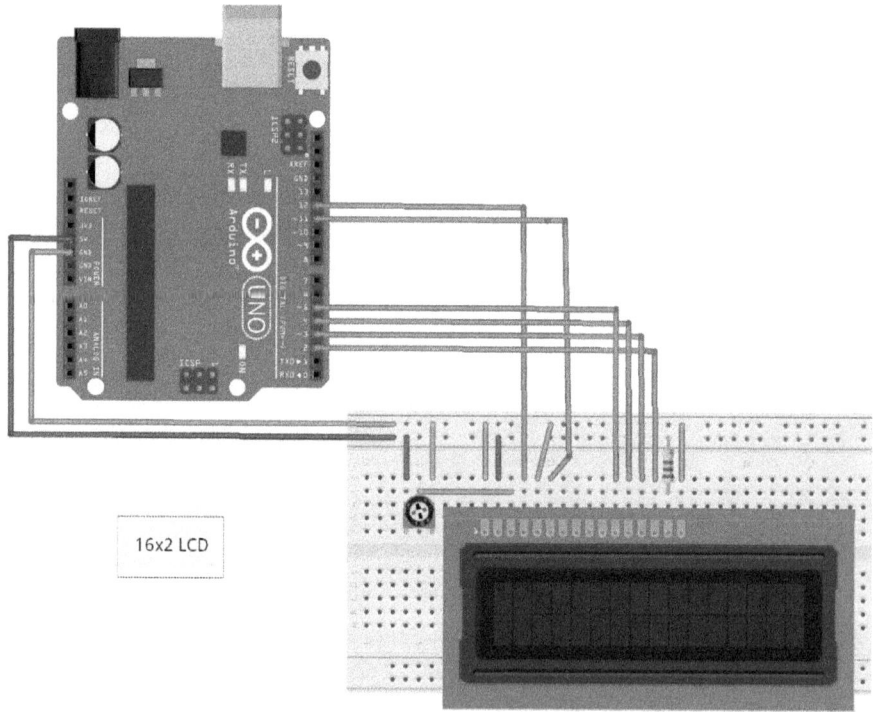

Fig 8.1 16x2 LCD Display Circuit Diagram

In our program, we will need to use the LiquidCrystal library. This allows us to control LCD displays that are compatible with the Hitachi HD44780 driver. The program below is quite straight forward and will display the words "Hello World" and the time in seconds since the Arduino was last reset on the LCD.

(This program also available from http://robertthomasson.blogspot.co.uk/2013/09/byte-size-starter.html):

```
#include <LiquidCrystal.h>        // include the library code

// initialize the library with the numbers of the interface pins
LiquidCrystal lcd(12, 11, 5, 4, 3, 2);

void setup() {

  lcd.begin(16, 2);               // set up the LCD's number of columns and rows
  lcd.print("hello, world!");     // Print a message to the LCD

}

void loop() {

  lcd.setCursor(0, 1);            // set the cursor to column 0, line 1
  lcd.print(millis()/1000);       // print the number of seconds since reset

}
```

The code is based on the Liquid Crystal example by Tom Igoe which can be found on the Arduino website at http://arduino.cc/en/Tutorial/LiquidCrystal

Note that the cursor is set to line 1 for printing of the number of seconds. Line 1 is the second row, as line 0 is the first row.

We now have an easy way of displaying text on the 16x2 LCD. Try altering the text or getting it to display temperature data. The latter will be covered in a future publication.

8.2 NOKIA 5110 LCD

The Nokia 5110 84x48 LCD graphics display was used on millions of mobile phones in the late 1990's and due to their simple connections they are easily integrated into electronics projects. They are inexpensive, versatile and are very reliable. They are available "as is" or can be purchased with header pins already attached.

Fig 8.1 Nokia 5110 LCD Module

Follow the pin out scheme below. Pins on bottom, screen face up, looking left to right.

Nokia 5110	Arduino
RST	6
CE	7
DC	5
DIN	4
CLK	3
VCC	3V
LIGHT	GND (for the backlight)
GND	GND

Now, upload the following code (rather than type this in, you can get it from the Arduino website at http://playground.arduino.cc/Code/PCD8544 or from http://robertthomasson.blogspot.co.uk/2013/09/byte-size-starter.html):

```
#define PIN_SCE   7
#define PIN_RESET 6
#define PIN_DC    5
#define PIN_SDIN  4
#define PIN_SCLK  3

#define LCD_C     LOW
#define LCD_D     HIGH
```

```
#define LCD_X    84
#define LCD_Y    48

static const byte ASCII[][5] =
{
 {0x00, 0x00, 0x00, 0x00, 0x00} // 20
,{0x00, 0x00, 0x5f, 0x00, 0x00} // 21 !
,{0x00, 0x07, 0x00, 0x07, 0x00} // 22 "
,{0x14, 0x7f, 0x14, 0x7f, 0x14} // 23 #
,{0x24, 0x2a, 0x7f, 0x2a, 0x12} // 24 $
,{0x23, 0x13, 0x08, 0x64, 0x62} // 25 %
,{0x36, 0x49, 0x55, 0x22, 0x50} // 26 &
,{0x00, 0x05, 0x03, 0x00, 0x00} // 27 '
,{0x00, 0x1c, 0x22, 0x41, 0x00} // 28 (
,{0x00, 0x41, 0x22, 0x1c, 0x00} // 29 )
,{0x14, 0x08, 0x3e, 0x08, 0x14} // 2a *
,{0x08, 0x08, 0x3e, 0x08, 0x08} // 2b +
,{0x00, 0x50, 0x30, 0x00, 0x00} // 2c ,
,{0x08, 0x08, 0x08, 0x08, 0x08} // 2d -
,{0x00, 0x60, 0x60, 0x00, 0x00} // 2e .
,{0x20, 0x10, 0x08, 0x04, 0x02} // 2f /
,{0x3e, 0x51, 0x49, 0x45, 0x3e} // 30 0
,{0x00, 0x42, 0x7f, 0x40, 0x00} // 31 1
,{0x42, 0x61, 0x51, 0x49, 0x46} // 32 2
,{0x21, 0x41, 0x45, 0x4b, 0x31} // 33 3
,{0x18, 0x14, 0x12, 0x7f, 0x10} // 34 4
,{0x27, 0x45, 0x45, 0x45, 0x39} // 35 5
,{0x3c, 0x4a, 0x49, 0x49, 0x30} // 36 6
,{0x01, 0x71, 0x09, 0x05, 0x03} // 37 7
,{0x36, 0x49, 0x49, 0x49, 0x36} // 38 8
,{0x06, 0x49, 0x49, 0x29, 0x1e} // 39 9
,{0x00, 0x36, 0x36, 0x00, 0x00} // 3a :
,{0x00, 0x56, 0x36, 0x00, 0x00} // 3b ;
,{0x08, 0x14, 0x22, 0x41, 0x00} // 3c <
,{0x14, 0x14, 0x14, 0x14, 0x14} // 3d =
,{0x00, 0x41, 0x22, 0x14, 0x08} // 3e >
,{0x02, 0x01, 0x51, 0x09, 0x06} // 3f ?
,{0x32, 0x49, 0x79, 0x41, 0x3e} // 40 @
,{0x7e, 0x11, 0x11, 0x11, 0x7e} // 41 A
,{0x7f, 0x49, 0x49, 0x49, 0x36} // 42 B
,{0x3e, 0x41, 0x41, 0x41, 0x22} // 43 C
,{0x7f, 0x41, 0x41, 0x22, 0x1c} // 44 D
,{0x7f, 0x49, 0x49, 0x49, 0x41} // 45 E
,{0x7f, 0x09, 0x09, 0x09, 0x01} // 46 F
```

,{0x3e, 0x41, 0x49, 0x49, 0x7a} // 47 G
,{0x7f, 0x08, 0x08, 0x08, 0x7f} // 48 H
,{0x00, 0x41, 0x7f, 0x41, 0x00} // 49 I
,{0x20, 0x40, 0x41, 0x3f, 0x01} // 4a J
,{0x7f, 0x08, 0x14, 0x22, 0x41} // 4b K
,{0x7f, 0x40, 0x40, 0x40, 0x40} // 4c L
,{0x7f, 0x02, 0x0c, 0x02, 0x7f} // 4d M
,{0x7f, 0x04, 0x08, 0x10, 0x7f} // 4e N
,{0x3e, 0x41, 0x41, 0x41, 0x3e} // 4f O
,{0x7f, 0x09, 0x09, 0x09, 0x06} // 50 P
,{0x3e, 0x41, 0x51, 0x21, 0x5e} // 51 Q
,{0x7f, 0x09, 0x19, 0x29, 0x46} // 52 R
,{0x46, 0x49, 0x49, 0x49, 0x31} // 53 S
,{0x01, 0x01, 0x7f, 0x01, 0x01} // 54 T
,{0x3f, 0x40, 0x40, 0x40, 0x3f} // 55 U
,{0x1f, 0x20, 0x40, 0x20, 0x1f} // 56 V
,{0x3f, 0x40, 0x38, 0x40, 0x3f} // 57 W
,{0x63, 0x14, 0x08, 0x14, 0x63} // 58 X
,{0x07, 0x08, 0x70, 0x08, 0x07} // 59 Y
,{0x61, 0x51, 0x49, 0x45, 0x43} // 5a Z
,{0x00, 0x7f, 0x41, 0x41, 0x00} // 5b [
,{0x02, 0x04, 0x08, 0x10, 0x20} // 5c ¥
,{0x00, 0x41, 0x41, 0x7f, 0x00} // 5d]
,{0x04, 0x02, 0x01, 0x02, 0x04} // 5e ^
,{0x40, 0x40, 0x40, 0x40, 0x40} // 5f _
,{0x00, 0x01, 0x02, 0x04, 0x00} // 60 `
,{0x20, 0x54, 0x54, 0x54, 0x78} // 61 a
,{0x7f, 0x48, 0x44, 0x44, 0x38} // 62 b
,{0x38, 0x44, 0x44, 0x44, 0x20} // 63 c
,{0x38, 0x44, 0x44, 0x48, 0x7f} // 64 d
,{0x38, 0x54, 0x54, 0x54, 0x18} // 65 e
,{0x08, 0x7e, 0x09, 0x01, 0x02} // 66 f
,{0x0c, 0x52, 0x52, 0x52, 0x3e} // 67 g
,{0x7f, 0x08, 0x04, 0x04, 0x78} // 68 h
,{0x00, 0x44, 0x7d, 0x40, 0x00} // 69 i
,{0x20, 0x40, 0x44, 0x3d, 0x00} // 6a j
,{0x7f, 0x10, 0x28, 0x44, 0x00} // 6b k
,{0x00, 0x41, 0x7f, 0x40, 0x00} // 6c l
,{0x7c, 0x04, 0x18, 0x04, 0x78} // 6d m
,{0x7c, 0x08, 0x04, 0x04, 0x78} // 6e n
,{0x38, 0x44, 0x44, 0x44, 0x38} // 6f o
,{0x7c, 0x14, 0x14, 0x14, 0x08} // 70 p
,{0x08, 0x14, 0x14, 0x18, 0x7c} // 71 q
,{0x7c, 0x08, 0x04, 0x04, 0x08} // 72 r
,{0x48, 0x54, 0x54, 0x54, 0x20} // 73 s
,{0x04, 0x3f, 0x44, 0x40, 0x20} // 74 t

```
,{0x3c, 0x40, 0x40, 0x20, 0x7c} // 75 u
,{0x1c, 0x20, 0x40, 0x20, 0x1c} // 76 v
,{0x3c, 0x40, 0x30, 0x40, 0x3c} // 77 w
,{0x44, 0x28, 0x10, 0x28, 0x44} // 78 x
,{0x0c, 0x50, 0x50, 0x50, 0x3c} // 79 y
,{0x44, 0x64, 0x54, 0x4c, 0x44} // 7a z
,{0x00, 0x08, 0x36, 0x41, 0x00} // 7b {
,{0x00, 0x00, 0x7f, 0x00, 0x00} // 7c |
,{0x00, 0x41, 0x36, 0x08, 0x00} // 7d }
,{0x10, 0x08, 0x08, 0x10, 0x08} // 7e ←
,{0x78, 0x46, 0x41, 0x46, 0x78} // 7f →
};

void LcdCharacter(char character)
{
  LcdWrite(LCD_D, 0x00);
  for (int index = 0; index < 5; index++)
  {
    LcdWrite(LCD_D, ASCII[character - 0x20][index]);
  }
  LcdWrite(LCD_D, 0x00);
}

void LcdClear(void)
{
  for (int index = 0; index < LCD_X * LCD_Y / 8; index++)
  {
    LcdWrite(LCD_D, 0x00);
  }
}

void LcdInitialise(void)
{
  pinMode(PIN_SCE, OUTPUT);
  pinMode(PIN_RESET, OUTPUT);
  pinMode(PIN_DC, OUTPUT);
  pinMode(PIN_SDIN, OUTPUT);
  pinMode(PIN_SCLK, OUTPUT);
  digitalWrite(PIN_RESET, LOW);
  digitalWrite(PIN_RESET, HIGH);
  LcdWrite(LCD_C, 0x21 );  // LCD Extended Commands.
  LcdWrite(LCD_C, 0xB1 );  // Set LCD Vop (Contrast).
  LcdWrite(LCD_C, 0x04 );  // Set Temp coefficient. //0x04
  LcdWrite(LCD_C, 0x14 );  // LCD bias mode 1:48. //0x13
  LcdWrite(LCD_C, 0x0C );  // LCD in normal mode.
```

```
  LcdWrite(LCD_C, 0x20 );
  LcdWrite(LCD_C, 0x0C );
}

void LcdString(char *characters)
{
  while (*characters)
  {
    LcdCharacter(*characters++);
  }
}

void LcdWrite(byte dc, byte data)
{
  digitalWrite(PIN_DC, dc);
  digitalWrite(PIN_SCE, LOW);
  shiftOut(PIN_SDIN, PIN_SCLK, MSBFIRST, data);
  digitalWrite(PIN_SCE, HIGH);
}

void setup(void)
{
  LcdInitialise();
  LcdClear();
  LcdString("Hello World!");
}

void loop(void)
{
}
```

There are plenty more examples freely available on the internet that will enable you to display text effects such as scrolling text as well as examples for displaying graphics.

Now we have learned how we can hook up and use two types of LCD with the Arduino. Integrating these LCDs with other components will be covered in future publications.

9 SUMMARY

I hope that you have enjoyed delving into the world of the Arduino microcontroller and that I have whetted your appetite to develop the basic ideas that were introduced to you.

In future publications, I will be introducing further components and integrating them with each other. These small projects will then be developed into larger projects that can be used and customised to be used in all manner of situations. In the meantime, keep experimenting by altering the programs and try to start integrating the topics yourself. Why not monitor humidity and display it on an LCD, hook up some buttons and a piezoelectric speaker to make a basic musical keypad or get a buzzer to alarm when a room gets dark?

There are no limits to what you can try and hopefully this book has inspired you to learn more and to experiment further.

ABOUT THE AUTHOR

Robert Thomasson is a research scientist and computer industry professional. Curious about anything scientific or technical, he was intrigued by the Arduino platform and the possible applications that it could be used in. He took up the challenge of learning about the hardware, software and peripheral components and how they can be adapted and used in everyday life situations. He wants to share his experiences and knowledge with anyone who will listen and to a wider audience as possible. He feels that anyone can learn about these things and with a basic knowledge can have a hobby for life or even build a career. He also feels it invaluable that younger people should learn about such things as they are the inventors and innovators of the future.

www.ingramcontent.com/pod-product-compliance
Lightning Source LLC
Chambersburg PA
CBHW071631170526
45166CB00003B/1291